Oklahoma images/covers/$3_{practice_tests}.jpeg4.3in4.3in065NavyBlue$

3 Oklahoma OSTP Grade 3 Math Practice Tests

Full-Length Test Prep with Detailed Answer Explanations

Dr. A. Nazari

3 Practice Tests to Get You Started!

Hey there, future math whiz! 

This book has **3 full practice tests** to help you warm up for the real thing. Think of it like stretching before a big game — these tests will get your brain ready and show you what to expect!

👍 Three tests is the **perfect start**!

👍 Each one helps you feel **more ready**!

👍 You'll be surprised how much you **already know**!

Sharpen your pencil and let's get warmed up! 

> 66 Three practice tests is a great way to start. Take your time with each one, and you'll feel more confident every step of the way! 99

📘 How to Use This Book 📘

📋 What's Inside This Book

- **3 Full-Length Practice Tests** — Each one covers all the Grade 3 math topics you need to know!
- **Answer Key with Explanations** — Find out why each answer is correct, not just what the answer is.
- **Reference Pages** — A math symbols chart and multiplication table you can peek at any time.
- **A Test Tracker** — Write down your scores and watch your confidence grow!

📅 A Simple 3-Test Plan

With just 3 tests, here's a great way to use them:

- **Test 1** — **The Warm-Up.** Take this test without a timer. Get comfortable with the question types. Don't worry about your score — just do your best!
- **Test 2** — **The Practice Round.** After reviewing Test 1, try this one with a timer (ask a grown-up!). Focus on the topics that were tricky last time.
- **Test 3** — **The Real Deal.** Treat this like the actual test: quiet room, timed, no peeking at answers. See how much you've improved!

⚪ Multiple Choice

Pick the **one best answer** from choices A, B, C, or D. Not sure? Cross out the ones you know are wrong, then pick from what's left. That's a smart move!

✏️ Short Answer

Write your answer **and** show your work! Even if your final answer isn't right, showing your steps can earn you credit. Use scratch paper if you need more room.

66 After Each Test 99

Flip to the Answer Key and check your work. For every question you got wrong, **read the explanation carefully**. Then write the tricky topics on your Test Tracker page. If you need extra help, grab our **Grade 3 Math Study Guide**!

> ⭐ **Fun fact**: Three tests is all it takes to see real improvement! Most kids feel way more confident after just a few rounds of practice. ⭐

Find more at
ViewMath.com/OK-Grade3

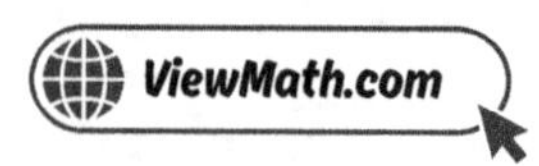

💡 Tips for Test Day 💡

Easy tricks to help you feel calm and do your best!

🌙 The Night Before

- ✅ **Sleep early** — your brain learns while you sleep!
- ✅ **Pack your supplies** — pencils, eraser, scratch paper, all ready to go.
- ✅ **Tell yourself:** "I've been practicing. I'm going to do great!"

👍 5 Simple Rules for Every Test

1. **Read the question twice.** The first time to understand it. The second time to catch details.
2. **Show your work.** Write the steps down, even on scratch paper. It helps you think!
3. **Skip the hard ones.** Put a small star next to tricky questions and come back later. Answer the easy ones first!
4. **Never leave a blank.** For multiple choice, your best guess is better than no answer at all.
5. **Check your work.** Finished early? Go back and re-read your answers.

✅ Smart Moves

- Take a deep breath before you begin
- Underline key words in the question
- Use drawings or number lines to help
- Cross out wrong answers first
- Double-check addition and subtraction

❌ Traps to Avoid

- Rushing and not reading carefully
- Picking the first answer that "looks right"
- Forgetting to carry or borrow numbers
- Skipping a question permanently
- Panicking when you see a tough problem

> 💬 *Remember, the very first practice test is the hardest — not because the questions are harder, but because everything is new! By Test 3, you'll feel like a pro. Trust me!* 💬

Get Ready to Practice

Here's everything you need before you start!

Pencils

Sharpened and ready!

Eraser

Everyone makes mistakes!

Scratch Paper

For working things out

A Calm Spot

Somewhere quiet to focus

A Grown-Up

To help set a timer

A Can-Do Attitude

You've totally got this!

✓ Allowed During Tests

- Pencils and erasers
- Blank scratch paper
- The **reference pages** in this book
- A ruler (for measurement questions)

✗ Not Allowed

- Calculators
- Phones, tablets, or computers
- Help from anyone else
- Your study guide (save it for after!)

For Parents & Teachers

- With only 3 tests, **space them at least a week apart**. This gives time to review mistakes before trying the next one.
- Let your child take Test 1 untimed to build familiarity.
- After each test, go through the Answer Key together. Focus on **understanding the "why,"** not just the score.
- If a topic keeps tripping them up, review it in our **Grade 3 Math Study Guide** before the next practice test.
- Celebrate every bit of progress — even getting one more question right is a win!

X^1 Math Reference Sheet X^1

You may use this page during your practice tests!

Symbol	Name	What It Means	
$+$	Plus (Add)	Put numbers together.	$3 + 5 = 8$
$-$	Minus (Subtract)	Take away from a number.	$9 - 4 = 5$
$\times$	Times (Multiply)	Add equal groups.	$4 \times 3 = 12$
$\div$	Divide	Split into equal groups.	$12 \div 3 = 4$
$=$	Equals	Both sides are the same.	$2 + 3 = 5$
$>$	Greater Than	The left number is bigger.	$7 > 3$
$<$	Less Than	The left number is smaller.	$2 < 9$
$\frac{1}{2}$	Fraction Bar	Part of a whole.	$\frac{1}{2}$ means 1 out of 2 equal parts

📖 Key Math Words

- **Sum** — the answer when you add
- **Difference** — the answer when you subtract
- **Product** — the answer when you multiply
- **Quotient** — the answer when you divide
- **Factor** — a number you multiply
- **Array** — objects in rows and columns
- **Fraction** — a part of a whole

- **Numerator** — the top number in a fraction
- **Denominator** — the bottom number
- **Equation** — a math sentence with $=$
- **Estimate** — a smart guess, close to the real answer
- **Perimeter** — the distance around a shape
- **Area** — the space inside a shape
- **Rounding** — making a number simpler by going to the nearest ten or hundred

🔍 Word Problem Clue Words

- **Add** (+): in all, total, altogether, combined, sum, both, more

- **Subtract** (−): how many more, how many left, fewer, difference, remain

- **Multiply** (×): each, every, groups of, times, rows of, per

- **Divide** (÷): share equally, split, each group, how many groups, per

Find more at
ViewMath.com/OK-Grade3

⣿ Multiplication Table ⣿

×	1	2	3	4	5	6	7	8	9	10	11
1	1	2	3	4	5	6	7	8	9	10	11
2	2	4	6	8	10	12	14	16	18	20	22
3	3	6	9	12	15	18	21	24	27	30	33
4	4	8	12	16	20	24	28	32	36	40	44
5	5	10	15	20	25	30	35	40	45	50	55
6	6	12	18	24	30	36	42	48	54	60	66
7	7	14	21	28	35	42	49	56	63	70	77
8	8	16	24	32	40	48	56	64	72	80	88
9	9	18	27	36	45	54	63	72	81	90	99
10	10	20	30	40	50	60	70	80	90	100	110
11	11	22	33	44	55	66	77	88	99	110	121

💡 How to Use This Table

To find **4 × 7**:

1. Find **4** in the left column (blue).
2. Find **7** in the top row (blue).
3. Follow the row and column until they meet: the answer is **28**!

📈 My Confidence Tracker 📈

Record your scores below. You'll be amazed at your progress!

My name: ___________________________________

✅ Test	📅 Date	⭐ Score	😊 How I Feel
1		/	
2		/	
3		/	

💡 My Quick Reflection

The easiest topic for me was:

____

The trickiest topic for me was:

____

One thing I got better at from Test 1 to Test 3:

____

Next time I want to try:

____

You just finished 3 practice tests — that's awesome! Compare your first score to your last. I bet you'll see real improvement. Ready for more? Check out our 5-test or 7-test books for even more practice!

Continue Learning at ViewMath Academy!

For Parents, Teachers & Students

Great job on the practice tests! Want to keep improving? ViewMath Academy is your **free online companion** to this book.

- **Score Analyzer** — Enter your answers and instantly see which topics need more practice

- **Interactive Lessons** — Review the concepts behind each question with clear explanations

- **Adaptive Quizzes** — Practice your weak topics with questions that match your level

- **Progress Tracking** — See your mastery grow across all Grade 3 math topics

- **Personalized Dashboard** — A learning plan tailored just for you

Scan to visit ViewMath Academy

viewmath.com/academy

 Free to use · No downloads required · Works on any device

 Table of Contents

Here's what we'll explore together!

 Let's learn and have fun!

Practice Test 1

30 Questions

✏ Before You Start ✏

- ✓ **Read each question carefully** before choosing your answer.
- ✓ **Show your work** on scratch paper when you need to.
- ✓ **Skip hard questions** and come back to them later.
- ✓ **Check your answers** when you're done.
- ✓ **Take your time** — there's no rush!

⭐ You've Got This! ⭐

Do your best and show what you know!

1. *Write 5,600 in expanded form.*

Your Answer:

2. *Which number rounds to 500 when rounded to the nearest 100?*

(A) 428

(B) 449

(C) 462

(D) 551

3. *Which group contains only even numbers?*

(A) 12, 34, 57

(B) 20, 46, 88

(C) 31, 50, 72

(D) 14, 63, 90

4. *A school raised $8,000 last year and $2,000 this year. How much did the school raise in total?*

(A) $6,000

(B) $10,000

(C) $82,000

(D) $28,000

5. *What is the value of the digit 8 in 283,041?*

(A) 80

(B) 800

(C) 8,000

(D) 80,000

6. *What is $198 + 345 + 267$?*

Your Answer:

Find more at
ViewMath.com/OK-Grade3

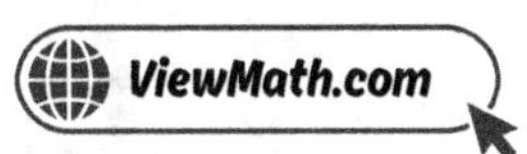

7. A store had 700 items in stock. After a sale, 263 items were sold. How many items remain?

Your Answer:

8. What is the missing digit? $8,\_\_46 - 3,289 = 4,757$

(A) 0   (B) 1

(C) 2   (D) 3

9. If $6 \times 9 = 54$, what is 9×6?

Your Answer:

10. What is $56 \div 8$?

(A) 6   (B) 7

(C) 8   (D) 48

11. A bakery makes 5 trays of muffins with 8 muffins each. Then they make 12 more muffins. How many muffins in all?

(A) 25   (B) 40

(C) 52   (D) 60

Find more at
ViewMath.com/OK-Grade3

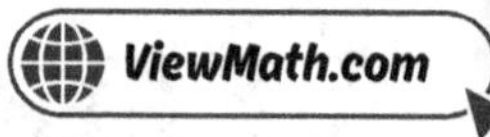

12. Look at this table:

Input	Output
12	4
18	6
?	8
30	10

What is the missing input?

(A) 20

(B) 22

(C) 24

(D) 26

---

13. The rule is "Add 9." If the input is 11, what is the output?

Your Answer:

---

14. Which fraction means "two-thirds"?

(A) $\frac{3}{2}$

(B) $\frac{2}{3}$

(C) $\frac{1}{3}$

(D) $\frac{2}{2}$

---

15. A number line is divided into 8 equal parts. How many marks from 0 is $\frac{7}{8}$?

Your Answer:

---

16. $\frac{7}{8}$ is built from how many copies of $\frac{1}{8}$?

(A) 1

(B) 7

(C) 8

(D) 15

Find more at
ViewMath.com/OK-Grade3

17. Which fraction is NOT equivalent to $\frac{2}{3}$?

  (A) $\frac{4}{6}$ (B) $\frac{6}{8}$

  (C) $\frac{6}{9}$ (D) $\frac{8}{12}$

18. Write 1 as a fraction with denominator 8.

Your Answer:

19. Compare: $\frac{3}{4}$ ___ $\frac{3}{8}$

  (A) $>$ (B) $<$

  (C) $=$ (D) Cannot compare

20. A string is $6\frac{1}{4}$ inches long. A ribbon is $6\frac{3}{4}$ inches long. How much longer is the ribbon than the string?

  (A) $\frac{1}{4}$ inch   (B) $\frac{1}{2}$ inch

  (C) $\frac{3}{4}$ inch   (D) 1 inch

21. A bag of apples has a mass of 3 kg. A bag of oranges has a mass of 5 kg. What is the total mass?

  (A) 2 kg   (B) 15 kg

  (C) 8 kg   (D) 35 kg

22. A large milk jug holds 4 liters. You pour out 1 liter for cereal. How much milk is left?

Your Answer:

Find more at
ViewMath.com/OK-Grade3

23. *How much is a quarter worth?*

(A) 1 cent

(B) 5 cents

(C) 10 cents

(D) 25 cents

24. *You buy a book for $4.50 and pay with a $10 bill. How much change do you get?*

(A) $4.50

(B) $5.50

(C) $6.50

(D) $14.50

25. *A bar graph has a scale that counts by 5s. A bar reaches up to the 4th line on the scale. What value does the bar show?*

(A) 4

(B) 9

(C) 15

(D) 20

26. *A line plot shows the lengths of 10 crayons. The data is shown below:*

2 in.	$2\frac{1}{2}$ in.	3 in.	$3\frac{1}{2}$ in.	4 in.
2	3	4	1	0

How many crayons are shorter than 3 inches?

(A) 2

(B) 4

(C) 5

(D) 9

27. *Is it possible to roll a number greater than 6 on a regular die numbered 1 through 6?*

Your Answer:

Find more at
ViewMath.com/OK-Grade3

28. A square has sides that are 7 inches long. What is its area?

   (A) 14 sq in

   (B) 28 sq in

   (C) 42 sq in

   (D) 49 sq in

29. Lily says a rectangle that is 5 cm long and 3 cm wide has a perimeter of 15 cm. What mistake did she make?

   (A) She subtracted instead of adding.

   (B) She found the area instead of the perimeter.

   (C) She forgot to add all 4 sides.

   (D) She divided instead of multiplying.

30. If one pizza is cut into 4 equal slices and another identical pizza is cut into 8 equal slices, which slices are bigger?

   (A) The $\frac{1}{8}$ slices

   (B) The $\frac{1}{4}$ slices

   (C) They are the same size.

   (D) You cannot tell.

Find more at
ViewMath.com/OK-Grade3

 # End of Practice Test 1 

Great job finishing the test!

 **My Score**

I got ____________ out of 30 questions right.

*Check your answers in the **Answer Key** at the back of the book.*

💡 *Review any questions you missed. That's how we learn!*

📊 Check Your Score Online!

Visit **ViewMath Academy** to enter your answers and see which topics you need to review. You can also explore lessons, take quizzes, track your scores, and save your progress!

viewmath.com/score/3.1.OK.01

Or go to viewmath.com/score and enter code: 3.1.OK.01

2

Practice Test 2

30 Questions

✏️ Before You Start ✏️

- ✓ **Read each question carefully** before choosing your answer.
- ✓ **Show your work** on scratch paper when you need to.
- ✓ **Skip hard questions** and come back to them later.
- ✓ **Check your answers** when you're done.
- ✓ **Take your time** — there's no rush!

⭐ You've Got This! ⭐

Do your best and show what you know!

1. Write the expanded form of 9,205.

   Your Answer:

2. Maria rounded 438 to the nearest 10 and got 430. Is she correct?

   (A) No, it should be 440

   (B) Yes, 438 rounds to 430

   (C) No, it should be 400

   (D) No, it should be 450

3. List all the even numbers between 21 and 30.

   Your Answer:

4. How many hundreds are in 10,000?

   Your Answer:

5. Which is the correct expanded form of 720,056?

   (A) $72,000 + 56$

   (B) $700,000 + 20,000 + 50 + 6$

   (C) $700,000 + 2,000 + 56$

   (D) $700,000 + 20,000 + 56$

6. A library has 583 fiction books and 268 nonfiction books. How many books does the library have in all?

   Your Answer:

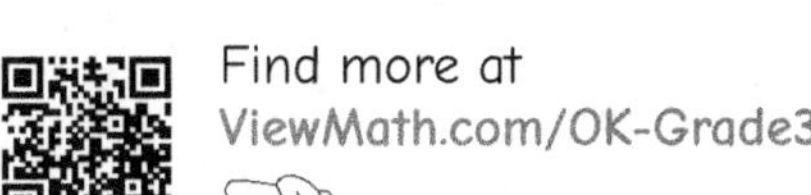

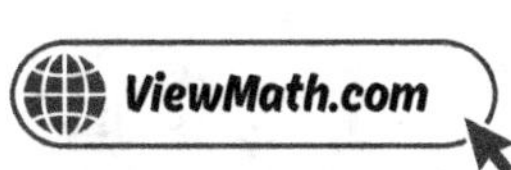

7. *What is the difference between 800 and 456?*

   (A) 344

   (B) 354

   (C) 444

   (D) 454

8. *There were 5,000 tickets for a concert. If 3,812 tickets were sold, how many are still available?*

   (A) 1,088

   (B) 1,188

   (C) 1,288

   (D) 2,188

9. *If $7 \times 9 = 63$, what is 9×7?*

   (A) 16

   (B) 56

   (C) 63

   (D) 79

10. *In the division sentence $48 \div 6 = 8$, what is the divisor?*

    (A) 48

    (B) 6

    (C) 8

    (D) 54

11. *Noah has $40. He buys 3 books that cost $7 each. How much money does he have left?*

    (A) $12

    (B) $19

    (C) $21

    (D) $33

12. *Which number is both a multiple of 4 AND a multiple of 6?*

    (A) 8

    (B) 12

    (C) 18

    (D) 16

Find more at
ViewMath.com/OK-Grade3

13. Look at the table. What is the missing output?

Input	Output
2	8
5	20
9	?

(A) 13

(B) 27

(C) 36

(D) 45

14. Lucy says $\frac{5}{4}$ means 5 parts out of 4. Is that possible?

(A) Yes, you just need more than one whole

(B) No, the numerator can never be bigger than the denominator

(C) No, fractions always equal less than 1

(D) Yes, but only with circles

15. A number line is divided into 4 equal parts. Which fraction is at the halfway point between 0 and 1?

(A) $\frac{1}{4}$

(B) $\frac{2}{4}$

(C) $\frac{3}{4}$

(D) $\frac{4}{4}$

16. How many copies of $\frac{1}{3}$ make 1 whole?

(A) 1

(B) 2

(C) 3

(D) 4

17. $\frac{3}{4} = \frac{?}{8}$. What is the missing numerator?

(A) 3

(B) 4

(C) 6

(D) 7

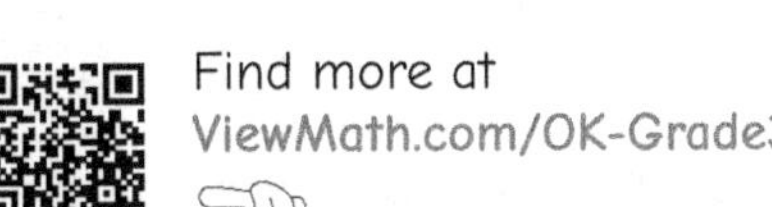
Find more at
ViewMath.com/OK-Grade3

18. *Write 4 as a fraction with denominator 8.*

Your Answer:

19. *Write <, >, or =:* $\frac{1}{4}$ ___ $\frac{1}{8}$

Your Answer:

20. *A marker reaches from 0 to the half-inch mark past 6 inches. How long is the marker?*

Your Answer:

21. *A melon has a mass of 2 kg. A bunch of grapes has a mass of 500 g. How much heavier is the melon than the grapes?*

(A) 498 g

(B) 1,000 g

(C) 1,500 g

(D) 2,500 g

22. *Container A holds 20 liters. Container B holds 13 liters. How much more does Container A hold?*

(A) 7 L

(B) 13 L

(C) 20 L

(D) 33 L

23. *Liam buys an apple for $0.85 and a banana for $0.40. He pays with a $5 bill. How much change does he get?*

(A) $3.25

(B) $3.75

(C) $4.15

(D) $4.60

24. *Subtract:* $12.00 − $7.45

25. Using the same rainy days graph (Jan = 6, Feb = 4, Mar = 3, Apr = 7), how many fewer rainy days were in February than in April?

(A) 1

(B) 2

(C) 3

(D) 4

26. Leo measured 8 worms. His line plot shows 7 X marks total. What went wrong?

(A) He measured one worm twice

(B) He forgot to plot one measurement

(C) He used the wrong numbers on the number line

(D) Nothing — 7 is correct

27. Is the following event certain, likely, unlikely, or impossible?
"Tomorrow will come after today."

28. A square has sides of 9 inches. What is the area?

29. A triangle has sides of 4 cm, 6 cm, and 8 cm. What is the perimeter?

(A) 14 cm

(B) 18 cm

(C) 20 cm

(D) 24 cm

Find more at
ViewMath.com/OK-Grade3

30. *You fold a piece of paper in half, then fold it in half again. How many equal parts do you have?*

Your Answer:

Find more at
ViewMath.com/OK-Grade3

 End of Practice Test 2 

Great job finishing the test!

My Score

I got _____________ out of 30 questions right.

*Check your answers in the **Answer Key** at the back of the book.*

Review any questions you missed. That's how we learn!

Check Your Score Online!

Visit **ViewMath Academy** to enter your answers and see which topics you need to review. You can also explore lessons, take quizzes, track your scores, and save your progress!

viewmath.com/score/3.1.OK.02

Or go to viewmath.com/score and enter code: 3.1.OK.02

Practice Test 3

 *30 Questions*

 Before You Start 

- ✓ **Read each question carefully** before choosing your answer.
- ✓ **Show your work** on scratch paper when you need to.
- ✓ **Skip hard questions** and come back to them later.
- ✓ **Check your answers** when you're done.
- ✓ **Take your time** — there's no rush!

 You've Got This! 

Do your best and show what you know!

1. Each place value is how many times bigger than the place to its right?

   (A) 2 times   (B) 5 times

   (C) 10 times   (D) 100 times

2. When you round 549 to the nearest 10 and to the nearest 100, which gives the larger answer?

   (A) Rounded to the nearest 10   (B) Rounded to the nearest 100

   (C) They are the same   (D) Cannot tell without rounding

3. If you add two odd numbers, the result is always:

   (A) Odd   (B) Even

   (C) Greater than 10   (D) An odd number less than 20

4. What number is 1 more than 9,999?

   (A) 9,000   (B) 9,990

   (C) 10,000   (D) 10,001

5. What is the smallest 6-digit number?

   Your Answer:

6. Which addition problem requires regrouping?

   (A) $135 + 243$ (B) $428 + 351$

   (C) $367 + 485$ (D) $512 + 136$

7. What is $986 - 543$?

(A) 433

(B) 443

(C) 453

(D) 543

8. A store had 6,250 items. They sold 3,487 items. How many items are left?

(A) 2,673

(B) 2,763

(C) 2,773

(D) 3,763

9. Which property is used here? $8 \times 7 = 8 \times (5 + 2) = (8 \times 5) + (8 \times 2)$

(A) Commutative property

(B) Associative property

(C) Distributive property

(D) Zero property

10. Carlos starts with 40 stickers. He gives away 5 stickers at a time. How many times can he give away stickers?

(A) 7

(B) 8

(C) 9

(D) 35

11. Mom bakes 36 cookies and divides them equally onto 4 plates. Dad adds 3 cookies to each plate. How many cookies are on each plate now?

(A) 10

(B) 11

(C) 12

(D) 13

12. What is the next number in this pattern? $7, 14, 21, 28, \ldots$

(A) 30

(B) 32

(C) 35

(D) 36

13. The rule is "Multiply by 6." The output is 48. What is the input?

(A) 6

(B) 7

(C) 8

(D) 9

14. Which fraction has a numerator of 2 and a denominator of 6?

(A) $\frac{6}{2}$

(B) $\frac{2}{6}$

(C) $\frac{2}{2}$

(D) $\frac{6}{6}$

15. Is $\frac{3}{3}$ at 0 or at 1 on the number line?

Your Answer:

16. On a number line, you are at $\frac{2}{8}$. You take 3 more hops of $\frac{1}{8}$. Where do you land?

(A) $\frac{3}{8}$

(B) $\frac{5}{8}$

(C) $\frac{5}{16}$

(D) $\frac{6}{8}$

17. Find a fraction equivalent to $\frac{1}{3}$ by multiplying the numerator and denominator by 2.

(A) $\frac{2}{3}$

(B) $\frac{1}{6}$

(C) $\frac{2}{6}$

(D) $\frac{3}{6}$

Find more at
ViewMath.com/OK-Grade3

18. What whole number does $\frac{6}{3}$ equal?

(A) 1

(B) 2

(C) 3

(D) 6

19. When two fractions have the same numerator but different denominators, which fraction is greater?

(A) The one with the bigger denominator

(B) The one with the smaller denominator

(C) They are always equal

(D) You cannot compare them

20. A ribbon is $5\frac{1}{2}$ inches long. A piece of string is $4\frac{3}{4}$ inches long. Which is longer?

(A) The ribbon

(B) The string

(C) They are the same length

(D) Not enough information

21. Convert 3 kilograms to grams.

Your Answer:

22. What unit do we use to measure liquid volume?

(A) Grams

(B) Kilograms

(C) Liters

(D) Inches

23. How many cents are in $1.00?

(A) 1 cent

(B) 10 cents

(C) 50 cents

(D) 100 cents

Find more at
ViewMath.com/OK-Grade3

24. What is $4.60 + $3.85?

(A) $7.45   (B) $8.45

(C) $8.35   (D) $7.85

25. A bar graph has a scale that counts by 10s. The bar for Red reaches the 3rd line. How many items does Red show?

Your Answer

26. Which type of graph is BEST for showing measurement data like the lengths of worms in inches?

(A) Picture graph   (B) Bar graph

(C) Line plot   (D) Tally chart

27. Which event is certain?

(A) You will roll a 3 on a die.   (B) The day after Monday is Tuesday.

(C) It will snow tomorrow.   (D) You will pick a red card from a deck.

28. A rectangle is 10 feet long and 3 feet wide. What is its area?

(A) 13 sq ft   (B) 26 sq ft

(C) 30 sq ft   (D) 33 sq ft

29. Which shape has the greatest perimeter?

(A) A square with sides of 5 cm   (B) A rectangle that is 8 cm × 3 cm

(C) A rectangle that is 6 cm × 4 cm   (D) A triangle with sides 7 cm, 7 cm, and 7 cm

Find more at
ViewMath.com/OK-Grade3

30. *A square is partitioned into 4 equal parts. All 4 parts are shaded. What fraction is shaded?*

(A) $\frac{1}{4}$

(B) $\frac{3}{4}$

(C) $\frac{4}{4}$

(D) $\frac{4}{1}$

Find more at
ViewMath.com/OK-Grade3

⭐ *End of Practice Test 3* ⭐

Great job finishing the test!

 *My Score*

I got _____________ out of 30 questions right.

*Check your answers in the **Answer Key** at the back of the book.*

 *Review any questions you missed. That's how we learn!*

📊 *Check Your Score Online!*

*Visit **ViewMath Academy** to enter your answers and see which topics you need to review. You can also explore lessons, take quizzes, track your scores, and save your progress!*

viewmath.com/score/3.1.OK.03

Or go to viewmath.com/score and enter code: 3.1.OK.03

Answer Key & Explanations

 **Check Your Answers!** 

First try each test on your own, then look here to check.

Read the explanations to learn from any mistakes ⭐

Practice Test 1 — Answer Key

1 $5{,}000 + 600$	**2** C	**3** B

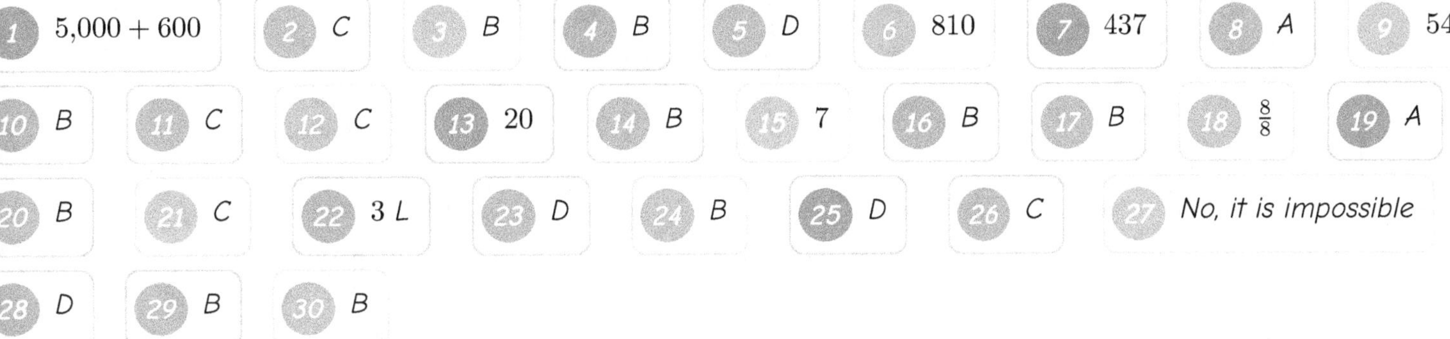

1 $5{,}000 + 600$ **2** C **3** B **4** B **5** D **6** 810 **7** 437 **8** A **9** 54

10 B **11** C **12** C **13** 20 **14** B **15** 7 **16** B **17** B **18** $\frac{8}{8}$ **19** A

20 B **21** C **22** 3 L **23** D **24** B **25** D **26** C **27** No, it is impossible

28 D **29** B **30** B

💡 **Time to Learn!** 💡

Go through the explanations below, **especially for the questions you missed.**

Understanding why each answer is correct makes you a stronger math thinker!

👍 **Tip:** Circle any questions you got wrong, then read their explanation carefully.

Practice Test 1 — Detailed Explanations

1 $5{,}600 = 5{,}000 + 600 + 0 + 0.$ *The tens and ones are both 0.*

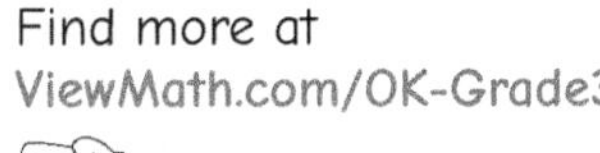

2 462 has tens digit $6 \geq 5$, so it rounds up to 500. 428 and 449 round to 400. 551 rounds to 600.

3 20 (ends in 0), 46 (ends in 6), 88 (ends in 8) are all even. The other groups each contain at least one odd number.

4 $8{,}000 + 2{,}000 = 10{,}000$.

5 In 283,041, the digit 8 is in the ten-thousands place, so its value is 80,000.

6 First add $198 + 345 = 543$. Ones: $8 + 5 = 13$, carry 1. Tens: $9 + 4 + 1 = 14$, carry 1. Hundreds: $1 + 3 + 1 = 5$. Then $543 + 267 = 810$. Ones: $3 + 7 = 10$, carry 1. Tens: $4 + 6 + 1 = 11$, carry 1. Hundreds: $5 + 2 + 1 = 8$.

7 $700 - 263$: borrow through the zeros. Ones: $10 - 3 = 7$. Tens: $9 - 6 = 3$. Hundreds: $6 - 2 = 4$. The store has 437 items remaining.

8 Add the difference and the subtracted number. $4{,}757 + 3{,}289 = 8{,}046$. The missing digit is 0.

9 By the commutative property, swapping the factors does not change the product.

10 $56 \div 8 = 7$ because $8 \times 7 = 56$.

11 Step 1: $5 \times 8 = 40$ muffins. Step 2: $40 + 12 = 52$ muffins in all.

12 The rule is "divide by 3": $12 \div 3 = 4$, $18 \div 3 = 6$, $30 \div 3 = 10$. Missing: $? \div 3 = 8$, so $? = 24$.

13 $11 + 9 = 20$.

14 "Two-thirds" means 2 parts out of 3 equal parts $= \frac{2}{3}$.

15 $\frac{7}{8}$ is 7 marks from 0 because the numerator counts the marks.

16 $\frac{7}{8} = 7$ copies of $\frac{1}{8}$.

17 $\frac{4}{6} = \frac{2}{3}$, $\frac{6}{9} = \frac{2}{3}$, and $\frac{8}{12} = \frac{2}{3}$. But $\frac{6}{8} = \frac{3}{4}$, which is not equivalent to $\frac{2}{3}$.

18 $1 = \frac{8}{8}$ because 8 eighths make 1 whole.

19   Same numerator (3 pieces). Fourths are bigger than eighths, so $\frac{3}{4} > \frac{3}{8}$.

20 $6\frac{3}{4} - 6\frac{1}{4} = \frac{3}{4} - \frac{1}{4} = \frac{2}{4} = \frac{1}{2}$ inch.

21 $3 + 5 = 8$ kg.

22 $4 - 1 = 3$ L.

23   A quarter is worth 25 cents.

24 $\$10.00 - \$4.50 = \$5.50$.

25   The scale counts by 5s, so the 4th line is $4 \times 5 = 20$.

26   Crayons shorter than 3 inches: 2 in. (2) $+ 2\frac{1}{2}$ in. (3) $= 5$ crayons.

27   A regular die only has numbers 1 through 6, so rolling a number greater than 6 is impossible.

28   A square is a rectangle with all sides equal. Area $= 7 \times 7 = 49$ sq in.

(29) Lily multiplied $5 \times 3 = 15$, which gives the area, not the perimeter. The correct perimeter is $5 + 3 + 5 + 3 = 16$ cm.

(30) The $\frac{1}{4}$ slices are bigger because the pizza is cut into fewer pieces. More equal parts means smaller pieces: $\frac{1}{4} > \frac{1}{8}$.

☑ Practice Test 2 — Answer Key

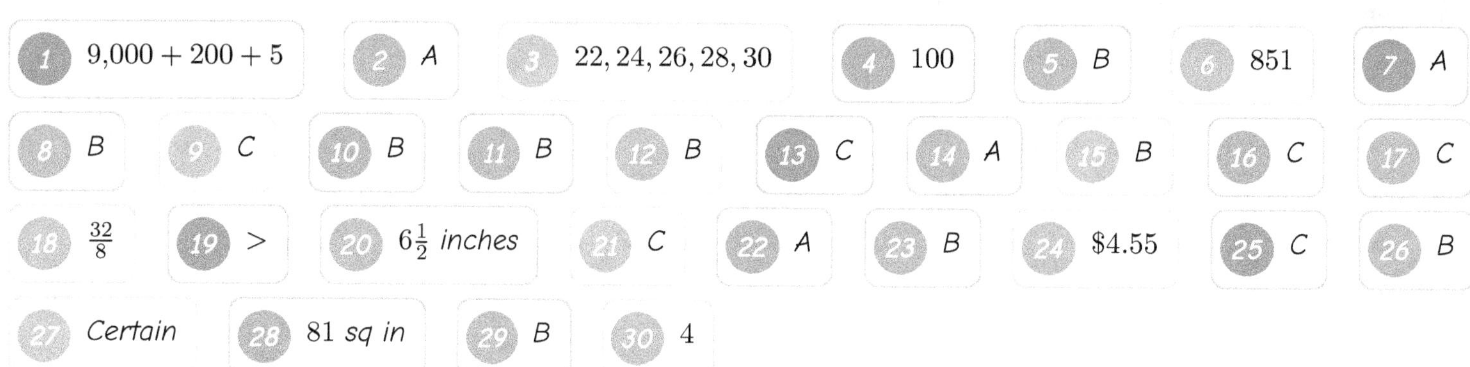

(1) $9,000 + 200 + 5$	(2) A	(3) $22, 24, 26, 28, 30$ (4) 100 (5) B (6) 851 (7) A
(8) B (9) C (10) B (11) B (12) B (13) C (14) A (15) B (16) C (17) C		
(18) $\frac{32}{8}$ (19) > (20) $6\frac{1}{2}$ inches (21) C (22) A (23) B (24) \$4.55 (25) C (26) B		
(27) Certain (28) 81 sq in (29) B (30) 4		

💡 Time to Learn! 💡

Go through the explanations below, **especially for the questions you missed**.

Understanding why each answer is correct makes you a stronger math thinker!

👍 **Tip:** Circle any questions you got wrong, then read their explanation carefully.

📖 Practice Test 2 — Detailed Explanations

(1) $9,205 = 9,000 + 200 + 0 + 5$. The tens digit is 0, so there is no tens term.

(2) The ones digit is 8. Since $8 \geq 5$, we round up. 438 rounds to 440, not 430.

Find more at
ViewMath.com/OK-Grade3

3. The even numbers between 21 and 30 are $22, 24, 26, 28, 30$. They all end in an even digit.

4. $10,000 \div 100 = 100$. There are 100 hundreds in 10,000.

5. $720,056 = 700,000 + 20,000 + 0 + 0 + 50 + 6$. The thousands and hundreds places are both zero.

6. $583 + 268 = 851$. Ones: $3 + 8 = 11$, carry 1. Tens: $8 + 6 + 1 = 15$, carry 1. Hundreds: $5 + 2 + 1 = 8$.

7. $800 - 456$: borrow through zeros. Ones: $10 - 6 = 4$. Tens: $9 - 5 = 4$. Hundreds: $7 - 4 = 3$. The difference is 344.

8. $5,000 - 3,812 = 1,188$. Borrow through zeros: $10 - 2 = 8$, $9 - 1 = 8$, $9 - 8 = 1$, $4 - 3 = 1$.

9. By the commutative property, $9 \times 7 = 7 \times 9 = 63$.

10. The divisor is the number you divide by. In $48 \div 6 = 8$, the divisor is 6.

11. Step 1: $3 \times \$7 = \$21$ spent. Step 2: $\$40 - \$21 = \$19$ left.

12. Multiples of 4: $4, 8, 12, 16, \ldots$ Multiples of 6: $6, 12, 18, \ldots$ The number 12 appears in both lists.

13. The rule is multiply by 4 ($2 \times 4 = 8$, $5 \times 4 = 20$). So $9 \times 4 = 36$.

14. $\frac{5}{4}$ means 5 parts where each part is $\frac{1}{4}$. That is more than one whole ($1\frac{1}{4}$).

15. $\frac{2}{4}$ is at the second of 4 marks, which is the middle of the line from 0 to 1.

16. $\frac{1}{3} + \frac{1}{3} + \frac{1}{3} = \frac{3}{3} = 1$ whole. It takes 3 copies.

Find more at
ViewMath.com/OK-Grade3

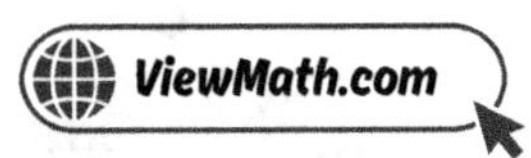

17 The denominator went from 4 to 8 (multiplied by 2). So multiply the numerator by 2 too: $3 \times 2 = 6$.

18 $4 \times 8 = 32$ eighths. So $4 = \frac{32}{8}$.

19 Same numerator. Fourths are bigger pieces than eighths. $\frac{1}{4} > \frac{1}{8}$.

20 The marker ends at the half-inch mark past 6, which is $6\frac{1}{2}$ inches.

21 $2 \ kg = 2{,}000 \ g$. Then $2{,}000 - 500 = 1{,}500 \ g$.

22 $20 - 13 = 7 \ L$.

23 Total: $\$0.85 + \$0.40 = \$1.25$. Change: $\$5.00 - \$1.25 = \$3.75$.

24 $\$12.00 - \$7.45 = \$4.55$. Borrow to subtract 45 cents from 0 cents.

25 $7 - 4 = 3$ fewer rainy days in February than April.

26 He measured 8 worms but only has 7 X marks, so he forgot to plot one measurement.

27 Tomorrow always comes after today. This will definitely happen.

28 Area $= 9 \times 9 = 81$ sq in.

29 Add all side lengths: $4 + 6 + 8 = 18 \ cm$.

30 Folding in half gives 2 parts. Folding in half again doubles it to 4 equal parts. Each part is $\frac{1}{4}$ of the whole.

Find more at
ViewMath.com/OK-Grade3

✅ Practice Test 3 — Answer Key

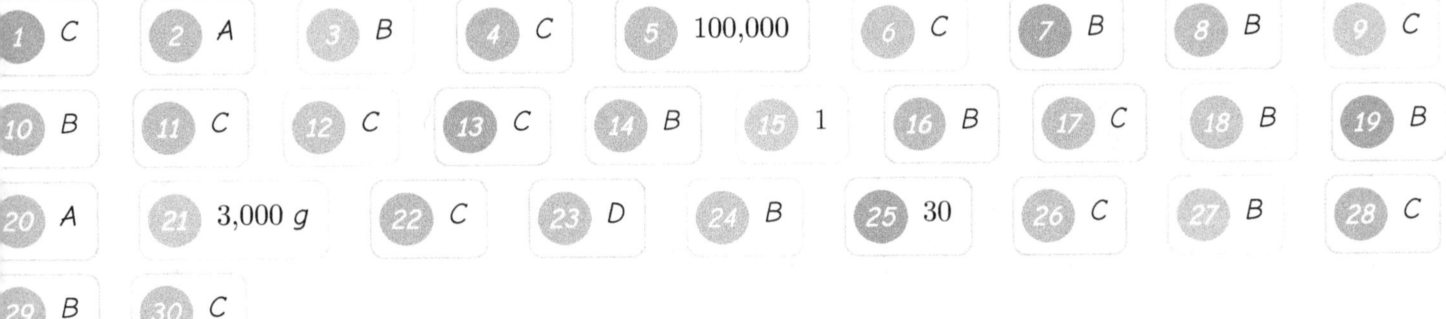

| 1 | C | 2 | A | 3 | B | 4 | C | 5 | 100,000 | 6 | C | 7 | B | 8 | B | 9 | C |

| 10 | B | 11 | C | 12 | C | 13 | C | 14 | B | 15 | 1 | 16 | B | 17 | C | 18 | B | 19 | B |

| 20 | A | 21 | 3,000 g | 22 | C | 23 | D | 24 | B | 25 | 30 | 26 | C | 27 | B | 28 | C |

| 29 | B | 30 | C |

💡 Time to Learn! 💡

Go through the explanations below, **especially for the questions you missed.**

Understanding why each answer is correct makes you a stronger math thinker!

👍 **Tip:** Circle any questions you got wrong, then read their explanation carefully.

📖 Practice Test 3 — Detailed Explanations

1 Each place is 10 times bigger than the place to its right: ones → tens → hundreds → thousands.

2 549 rounded to the nearest 10: ones digit is $9 \geq 5$, so 550. Rounded to the nearest 100: tens digit is $4 < 5$, so 500. $550 > 500$.

3 Odd + Odd = Even. For example, $3 + 5 = 8$ (even) and $7 + 9 = 16$ (even).

4 $9,999 + 1 = 10,000.$ This is the first 5-digit number.

Find more at
ViewMath.com/OK-Grade3

(5) The smallest 6-digit number is 100,000. Any number below it (like 99,999) has only 5 digits.

(6) In $367 + 485$, the ones column is $7 + 5 = 12$, which is 10 or more, so you must regroup. The other problems do not need regrouping.

(7) No borrowing is needed. Ones: $6 - 3 = 3$. Tens: $8 - 4 = 4$. Hundreds: $9 - 5 = 4$. The difference is 443.

(8) $6{,}250 - 3{,}487 = 2{,}763$. Borrow as needed in each column: ones $10 - 7 = 3$, tens $14 - 8 = 6$, hundreds $1 - 4$, borrow again: $11 - 4 = 7$, thousands $5 - 3 = 2$.

(9) Breaking apart a factor and multiplying each part is the distributive property.

(10) $40 \div 5 = 8$. He can give away stickers 8 times. This is like subtracting 5 from 40 repeatedly until reaching 0, which takes 8 steps.

(11) Step 1: $36 \div 4 = 9$ cookies per plate. Step 2: $9 + 3 = 12$ cookies per plate.

(12) The pattern adds 7 each time (multiples of 7): $28 + 7 = 35$.

(13) If the rule is multiply by 6 and the output is 48, then input $= 48 \div 6 = 8$.

(14) Numerator is on top and denominator is on the bottom. So it is $\frac{2}{6}$.

(15) $\frac{3}{3}$ means all 3 parts, which equals 1 whole.

(16) $\frac{2}{8} + 3$ hops of $\frac{1}{8} = \frac{2}{8} + \frac{3}{8} = \frac{5}{8}$.

(17) $\frac{1 \times 2}{3 \times 2} = \frac{2}{6}$. So $\frac{1}{3} = \frac{2}{6}$.

18 $\frac{3}{3} = 1$, so $\frac{6}{3} = 2$.

19   Same numerator means the same number of pieces. A smaller denominator means bigger pieces, so the fraction with the smaller denominator is greater.

20 $5\frac{1}{2} = 5\frac{2}{4}$, which is greater than $4\frac{3}{4}$. The ribbon is longer.

21 $3 \times 1{,}000 = 3{,}000$ g.

22   We measure liquid volume in liters (L).

23 $\$1.00 = 100$ cents.

24   Cents: $60 + 85 = 145$ cents $= 1$ dollar and 45 cents. Dollars: $4 + 3 + 1 = 8$. Answer: $\$8.45$.

25   The scale counts by 10s, so the 3rd line is $3 \times 10 = 30$ items.

26   Line plots are best for measurement data. They show each individual measurement as an X on a number line, making it easy to see how values are spread out.

27   Tuesday always comes after Monday. This will definitely happen, so it is certain.

28   Area $= 10 \times 3 = 30$ sq ft.

29   Square: $4 \times 5 = 20$ cm. Rectangle 8×3: $(2 \times 8) + (2 \times 3) = 22$ cm. Rectangle 6×4: $(2 \times 6) + (2 \times 4) = 20$ cm. Triangle: $7 + 7 + 7 = 21$ cm. The 8×3 rectangle has the greatest perimeter of 22 cm.

30   All 4 out of 4 parts are shaded, so $\frac{4}{4}$ is shaded. $\frac{4}{4} = 1$ whole.

Great job checking your work!

Keep practicing and you'll be a math star!